Light

Kay Davies
and
Wendy Oldfield

Wayland

Starting Science

Books in the series

Animals
Electricity and Magnetism
Floating and Sinking
Food
Hot and Cold
Information Technology

Light
Skeletons and Movement
Sound and Music
Waste
Water
Weather

About this book

Information in *Light* is closely linked to practical activities dealing with this subject, so that children can learn from first hand experience. Topics include how light is produced and how it is reflected and refracted. Children discover that light travels in straight lines and also how colours can be mixed. Investigations with colour and pattern show their importance in the natural world and how they can play tricks on our eyes.

This book provides an introduction to methods in scientific enquiry and recording. The activities and investigations are designed to be straightforward but fun, and flexible according to the abilities of the children.

The main picture and its commentary may be taken as an introduction to the topic or as a focal point for further discussion. Each chapter can form a basis for extended topic work.

Teachers will find that in using this book, they are reinforcing the other core subjects of language and mathematics. Through its topic approach *Light* covers aspects of the National Science Curriculum for key stage 1 (levels 1 to 3), for the following Attainment Targets: Exploration of science (AT 1), The variety of life (AT 2), Types and uses of materials (AT 6), and Using light and electromagnetic radiation (AT 15).

First published in 1991 by
Wayland (Publishers) Ltd
61 Western Road, Hove
East Sussex, BN3 1JD, England

© Copyright 1991 Wayland (Publishers) Ltd

Typeset by Nicola Taylor, Wayland
Printed in Italy by
Rotolito Lombarda S.p.A., Milan
Bound in Belgium by Casterman S.A.

British Library Cataloguing in Publication Data
Oldfield, Wendy
  Light.
  1. Light
  I. Title II. Davies, Kay *1946–* III. Series
  535

ISBN 1 85210 996 3

Editor: Cally Chambers

CONTENTS

Light shines from the glowing sun on to the earth.
We can see it reflected off the pale moon too.

LIGHTING UP THE SKY

The sun is a star. Stars are very hot and give out light.

The earth and moon do not give out light.

The light from the sun falls on the earth. It reflects or bounces off things. This is how we can see them.

When things burn or get really hot, they give off their own light.
Look at these things.

Sort them into two groups.
1. Those that make their own light when they get hot.
2. Those that have no light of their own but reflect light into our eyes.

BRIGHT EYES

We can only see a small part of an eye.
It is round like a ball and kept safe inside a bony head.

We can see the coloured
iris and dark pupil.

The pupil is an opening
which lets light shine into
your eye.

Bright light makes the
pupil small. This stops too
much light getting in.

In the dark, the pupil
opens wide to let in lots
of light so we can see
better.

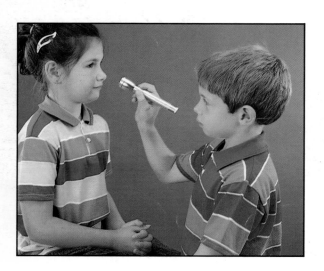

Gently shine a weak
torch into your
friend's eyes.

Watch their pupils.

What happens when you
switch the torch off?

It isn't easy to see in the dark forest at night.
The bushbaby has huge eyes to catch lots of light.

MIRROR MONSTERS

When light bounces off shiny things it gives a reflection that we can see.

Mirrors that are flat give the best reflections.

Curved mirrors can make things look bigger, smaller or change their shape. They can even turn things upside-down.

Try looking at your face in the front of a spoon and then the back.

Look at your reflection in a shiny car. What do you notice?

Look around your home and school for other things that make reflections.

Make a collection of shiny things that reflect light.

What are they made of?

Wavy mirrors twist and bend shapes.
We can see ourselves, but we look like strange monsters.

The flat surface of the lake is like a mirror.
Turn the picture upside-down. Does it look the same?

MIRROR IMAGE

If you look into a mirror you see a reflection of yourself.

This is your image.

Raise your right hand.

Does your image do the same?

Wink your left eye. Watch what your image does.

How is your image different from you?

Touch the surface of the mirror with the end of a ruler.

How far away from the mirror is your hand?

How far away from the mirror is your image's hand?

Where do you think the image really is?

BOUNCING AROUND

Mirrors can help you play tricks with light.

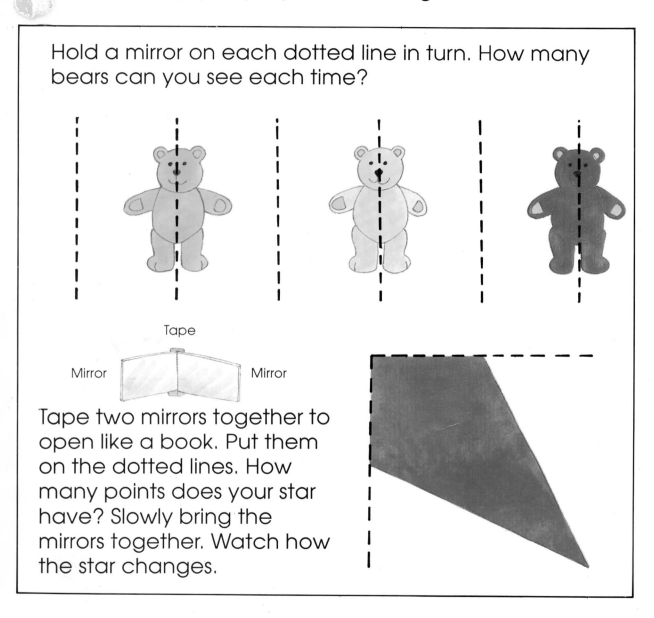

Hold a mirror on each dotted line in turn. How many bears can you see each time?

Tape

Mirror   Mirror

Tape two mirrors together to open like a book. Put them on the dotted lines. How many points does your star have? Slowly bring the mirrors together. Watch how the star changes.

You can make a kaleidoscope from three small mirrors. Tape them together to make a triangle. Drop paper shapes inside. Look into the corners to see your patterns.

Light bounces between the mirrors in a kaleidoscope.
It turns a handful of shapes into a beautiful pattern.

In strong light, shadows follow us everywhere.
They stand in our shoes and copy everything we do.

ME AND MY SHADOW

Light always travels in straight lines.
It cannot bend around things.

When an object blocks
the light's path, dark
shapes called shadows
appear.

Use your hands to make
some shadow shapes on
the wall.

You will need a bright
light behind you.

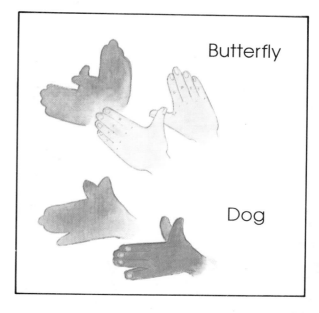

Butterfly

Dog

Use lots of things to make
shadows.
Move them about.

Do the shadows change
shape as you move the
objects?

Play a shadow guessing
game with your friends.

Do they know what is
making each shadow?

The sun shines through the mist and makes a rainbow.
Do you know the colours? Are they always the same?

SUNSHINE THROUGH RAIN

Things that we can see right through are transparent.

Air, water and glass are all transparent. Light can pass through them. Can you find anything transparent?

Usually the light from the sun or a lamp looks white. Light that looks white is really a mixture of many colours.

If sunlight shines through rain or mist, these colours can show as a rainbow.

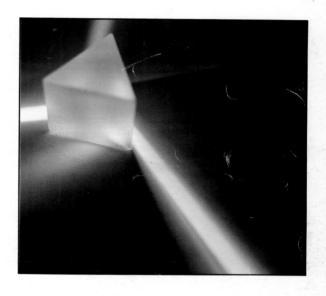

Glass, like this prism, can make rainbow colours too.

Look at the picture. How many colours can you find in light?

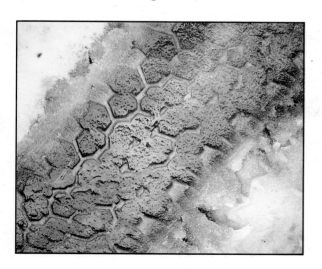

Look for colours in puddles and at the edges of mirrors, jewellery, cut glass and fish tanks.

Blow some bubbles. Can you find rainbow colours in them?

MIX AND MATCH

Put clean water in a **jar**.

Squeeze a few drops of red paint from a brush into the water.

Add some drops of blue paint to the jar.

What colour can you see where the paints mix? Have you made a new colour?

Paint a colour cake like this.

Paint three slices with just red, yellow and blue.

Mix these paints in pairs for the other three slices.

Can you name the new colours you have made?

What happens if you mix all three paints together?

The artist is painting a picture. He mixes paints together to make many different colours.

The window is made of coloured glass. The sun shines
through and makes patterns of light on the floor.

CHANGING COLOURS

Coloured light can make things look different.

Make a peep-hole in the front of a shoebox.

Make a larger hole in its lid. Cover this hole with a colour filter.

Put different coloured objects inside the box.

Hold the box in bright light and look through the peep-hole.

Try different colour filters.

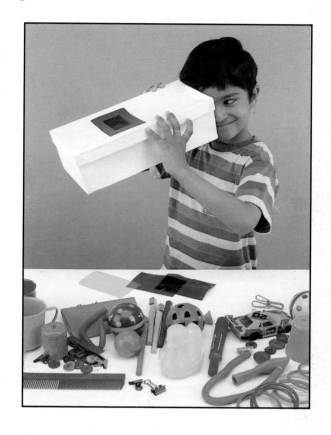

How do the colours in the box change each time?

Make a chart to record what you see.

Colour of light filter:	Red objects looked:	Green objects looked:	Blue objects looked:	Yellow objects looked:
Red	RED	BLACK	PURPLE	ORANGE
Green				
Blue				
Yellow				

The insect looks like the leaves it lives on. It is the same shape and colour. It can only be seen if it moves.

WHERE AM I?

Many creatures have colours and patterns which make them hard to see. They are camouflaged.

Their colouring and patterns are very like the plants or land on which they live. This helps them to hide from other creatures that might like to eat them.

Can you find the creatures hidden in this picture?

Make some snakes. Make some brightly coloured. Make some camouflaged. Stick them on to a background of grass and trees. Which are hardest to see?

I'M WARNING YOU!

Some pairs of colours really stand out.

Many poisonous creatures and plants use these colours to tell animals to leave them alone.

Make some colour cards like these. Stick different coloured crosses on backgrounds.

Make sure they are all the same size.

Pin all the cards on the wall.

Ask your friends to stand at the other side of the room.

Ask them which one they find easiest to see.

Make a block graph to show your results.

Has the wasp got his colours right?

The wasp doesn't need to hide. Its black and yellow bands warn other creatures that it is dangerous.

Familiar objects look strange in close-up.
Can you guess what this is? You might like to eat it.

CLOSE-UP

Lenses are specially shaped pieces of glass or other transparent material. Some bulge out and some curve in. They bend the light.

Magnifying glasses are lenses. Both sides of the lenses bulge out.

Use one to look at the details in a picture and at objects around you.
You might notice things you don't usually see.

Make your own lens.

Spread a piece of cling film over a roll of sticky tape. Drop a little water on to it.

Hold it over words in a newspaper.

Try adding more water and gently stretching the cling film to find the best magnifying glass.

PLAYING TRICKS

Things aren't always what they seem. Strong colours and patterns of lines can fool our eyes.

Look at the picture below. The bricklayers have built a wall with black and white bricks.
Have they laid them straight? Use a ruler to check.

Moving pictures trick us too. Make a little book.
Start at the back and copy the drawings in order.
Draw each one on a fresh page near the edge. Flick through the pages from back to front with your thumb.
Can you see the person moving?

Hot air rises from the baking desert.
It can play tricks with the light and we see mirages.

29

GLOSSARY

Camouflaged An animal is camouflaged when it is difficult to see because its colouring and pattern look like its surroundings.

Colour filter Transparent plastic or glass that lets through light only of the same colour as the filter.

Cut glass A kind of glass that has patterns cut into it.

Image The copy of something we see when we look in a mirror or other shiny surface.

Kaleidoscope A collection of mirrors that bounce light between them to make colourful patterns.

Lenses Pieces of transparent material with curved surfaces which bend light.

Magnifying glass A lens which makes objects look bigger.

Mirage An image of a lake which isn't really there.

Poisonous Animals and plants are poisonous if, by their bite or by being eaten, they cause illness or death to other creatures.

Prism A transparent triangle shape that can separate white light into rainbow colours.

Reflect To bounce light back off an object.

Reflection The image we see reflected in a shiny surface.

Transparent Materials, like glass or plastic, which let light pass through. We can see through transparent objects.

FINDING OUT MORE

Books to read:

Light (Argon and Xenon) by Ralph Chase (Blackwell Education, 1990)
Light by Angela Webb (Franklin Watts, 1988)
Look at Eyes by Henry Pluckrose (Franklin Watts, 1988)
My Mirror by Kay Davies and Wendy Oldfield (A & C Black, 1989)
My Shadow by Sheila Gore (A & C Black, 1989)
Your Eyes by Joan Iveson-Iveson (Wayland, 1985)

PICTURE ACKNOWLEDGEMENTS

Chris Fairclough Colour Library 19; Eye Ubiquitous 13; Hutchison 4; PHOTRI 26; Tony Stone Worldwide 10, 14, 16, 17 both; Wayland Picture Library 9, (Zul Mukhida) cover, 6 bottom, 8, 11 both, 15, 18 both, 21, 24, 27 both; ZEFA 6 top, 7, 20, 22, 25, 29.
Artwork illustrations by Rebecca Archer.
The publishers would also like to thank Davigdor Infants' School and Somerhill Road County Primary School, Hove, St Bernadette's First & Middle School and Downs County First School, Brighton, East Sussex, for their kind co-operation.

INDEX

Page numbers in bold indicate subjects shown in pictures, but not mentioned in the text on those pages.